IMAGES
of America

EARLY WARNER
BROS. STUDIOS

WARNER BROS., 1926. Portraits, from left to right, of Jack, Harry, Albert, and Sam Warner grace a picture of the Warner Bros. Sunset Studios in Hollywood. This was the first permanent home for the Warners, who had produced films in a series of rental stages before building here in 1920. The brothers had recently purchased Vitagraph Studios and were a year away from releasing *The Jazz Singer* (1927). (Courtesy Bison Archives.)

ON THE COVER: This aerial photograph of Warner Bros. First National Studios in Burbank, California, was taken on August 30, 1931, during Warner's fourth year in Burbank. In 1931, James Cagney's *The Public Enemy* and Edward G. Robinson's *Little Caesar* were released. The roles catapulted the men to stardom and began WB's run as the "Gangster Studio." (Courtesy Bison Archives.)

IMAGES
of America

EARLY WARNER
BROS. STUDIOS

E.J. Stephens and
Marc Wanamaker

ARCADIA
PUBLISHING

Published by Arcadia Publishing
Charleston, South Carolina

Library of Congress Control Number: 2010921991

For all general information contact Arcadia Publishing at:
Telephone 843-853-2070
Fax 843-853-0044
E-mail sales@arcadiapublishing.com
For customer service and orders:
Toll-Free 1-888-313-2665

Visit us on the Internet at www.arcadiapublishing.com

*This book is dedicated to the thousands of hardworking
Warner Bros. men and women who have been inspiring,
enlightening, and entertaining the world for nearly a century.*

CONTENTS

ACKNOWLEDGMENTS

It takes a lot of dedicated people to create a studio. The same can be said about creating a book about a studio. This book would have never been possible without the kind assistance of Chris Epting, who introduced E.J. to editor-extraordinaire Jerry Roberts, who in turn introduced E.J. to Marc. Thank you to Jerry Kern and Candy Gonzalez of the Warner Film Center at the Cascade Theatre, and Anna Mary Mooney of the Lawrence County (Pennsylvania) Historical Society, who keep the story of the Cascade Theatre in New Castle, Pennsylvania, alive. Two books proved invaluable in the writing of this book: Michael Freedland's *The Warner Brothers* and Cass Warner Spelling's *Hollywood Be Thy Name*. All images are courtesy of Bison Archives.

E.J. would like to thank his two beautiful kids, Mariah and Dylan, for putting up with his grumpiness as this book neared its deadline. He would especially like to thank his wife, best friend, and soul mate (who are all conveniently wrapped up into one beautiful woman named Kimi) for continually choosing him.

INTRODUCTION

If you wanted to concoct a mogul for one of Hollywood's major film factories during the early years of the motion picture industry, the formula generally went something like this: Take a group of impoverished, ambitious, first- or second-generation male Jewish immigrants from Eastern Europe, limit their formal education but school them in the competitive profession of sales in rugged eastern U.S. cities, give them the vision to glimpse the potential of motion pictures where others see only a fad, and most importantly, provide them the *chutzpah* to act on their insights.

This formula, with only slight modifications, produced the biographies of the founders of Universal, Fox, Columbia, MGM, and (the focus of this book) Warner Bros. The four Warner brothers—Harry, Albert, Sam, and Jack—were sons of a Polish cobbler and peddler, who moved his family to America to escape the ever-present threat of terror levied against Jews in his native land. Through vision, hard work, and the occasional lucky gamble, they were able to rise above "the huddled masses yearning to breathe free" to become Hollywood studio chiefs within a generation.

The meteoric rise of the Warner brothers is a tale so unlikely that had it been submitted as a screenplay at their own studio, it may well have been rejected as unbelievable. The story begins in 1903, at roughly the same time another set of siblings from the Midwest named Orville and Wilbur Wright were busy birthing the airplane. Nineteen-year-old Albert, the second-oldest Warner brother, was in Pittsburgh selling soap when he chanced upon a nickelodeon. He was so smitten by the hand-cranked silent films projected onto a stretched bed sheet that he returned to the Warner home in Youngstown, Ohio, with only one desire in mind: to get into the motion picture business. He was shocked to learn that his brothers Harry, 23, and Sam, 17, had come to the very same idea independently.

As fortune would have it, a friend offered to sell the boys an early projector called a Kinetoscope, along with a ragged copy of *The Great Train Robbery* (1903) for $1,000. The brothers, who added 12-year-old Jack to the venture, pooled their money, but came up $175 short. Their father saved the day when he tossed his gold watch and a family horse into the pot.

For the next several months the brothers traveled throughout Ohio and Pennsylvania showing their frayed film to miners in rented halls. They raised enough money during their tour, and from the sale of the family's bicycle shop, to open a nickelodeon of their own called the Cascade Theatre in New Castle, Pennsylvania. Since funds were always tight, they had to borrow chairs from a nearby mortuary, forever fearing that one of their shows would be disrupted by a well-attended funeral.

The brothers settled into roles at the Cascade that they would continue to play in one fashion or another for the rest of their careers. Harry, the oldest and most sober, handled the money; affable Sam, a technically minded dreamer, cranked the projector; salesman Albert hawked the tickets; and showman Jack "entertained" the crowds as the "chaser"—the less-than-gifted singer whose performances helped clear the theater between shows.

It did not take long for the brothers to see that the "reel money" to be made in motion pictures was to be found in distribution. This led them to create the Duquesne Amusement Supply Company. This film exchange, as distributors were known at the time, was a successful concern for several years until crushed in 1910 by Thomas Edison's monopolistic film trust.

Not ones to give up without a fight, the Warners scrounged up enough cash to move into film production on the East Coast. For the next few years, the fledgling enterprise was kept afloat by producing cliffhanger serials and a film made for the U.S. Army warning World War I soldiers about the dangers of venereal disease. In 1918, the brothers scored their first legitimate hit with

My Four Years in Germany, which funded a move to the West Coast, where they rented space in a studio near downtown Los Angeles. Within months they took up permanent residence on Hollywood's "Poverty Row."

By 1923, the brothers were successful enough to attract the talents of director Ernst Lubitsch, writer-producer Darryl F. Zanuck, and actor John Barrymore. But it was actually a four-legged star named Rin Tin Tin who kept the lights lit at Sunset and Bronson. Jack Warner liked to call him "The Mortgage Lifter."

The Warners, who always believed in expansion, took the biggest gamble of their careers in 1925 by purchasing Vitagraph Studios, one of filmdom's founding enterprises. This acquisition gave them two new studios, access to Vitagraph's vast film library, plus 50 new exchanges in North America and Europe.

On March 3, 1925, Warner Bros. launched KFWB, their first radio station in Los Angeles. Three of the four Warner brothers felt it should primarily function as a vehicle to promote Warner Bros. films. Sam Warner, however, saw the radio station as the first step in ushering in a new era of motion pictures where the audiences not only saw their heroes on the screen, but *heard* them as well. Sam had a lifelong fascination with machines and technology and worked closely with an engineer setting up KFWB's studios. That same engineer later showed him a process he was working on to synchronize sound to moving pictures. Further demonstrations convinced the other three brothers of the merits of the new system, which they purchased and renamed "Vitaphone" to capitalize on their recent acquisition of Vitagraph.

The following year, Warner Bros. produced *Don Juan*, starring John Barrymore, as their first full-length feature employing the Vitaphone process for music and sound effects. Sam oversaw the recording of an orchestra for the film in the Metropolitan Opera House in New York (a location that was problematic, as a subway line ran directly below the theater). *Don Juan* premiered at the Warners' Theatre in Manhattan on August 6, 1926, and all 1,200 seats sold out at the record price of $11 per ticket. In spite of the hugely successful premiere, the film was unable to recoup its costs, and the brothers were left seriously in debt.

Don Juan was WB's warning shot fired across the bow of silent films. The brothers effectively killed the genre 16 months later with their release of *The Jazz Singer* (1927), in which Al Jolson uttered the first words ever heard in a major Hollywood film: "Wait a minute, wait a minute . . . You ain't heard nothin' yet!"

The film was a smash hit. Seemingly overnight, theaters across the country were wired for sound, new soundstages were constructed, and foreign-accented actors by the dozen saw their studio contracts lapse. Tragically, the stress that came from the sound gamble lessened the Warner brothers' ranks by one. In a cruel twist of fate, Sam Warner, the brother most responsible for the company's—and Hollywood's—move into talkies, died from a cerebral hemorrhage 24 hours before *The Jazz Singer* premiered. He was only 42 years old.

By 1928, the three remaining brothers had become rich from spearheading the talkie movement but even so, they felt their futures were far from secure. They decided it was time for Warner Bros. to become a major studio. All they needed was cash, and Harry Warner, the financial genius of the family, was able to get his hands on piles of it from eager Wall Street bankers.

In a single deal, the Warners left Poverty Row behind forever by acquiring the huge but financially shaky First National Pictures company, which owned the largest theater chain in the country, as well as an enormous new studio complex a few miles across the hills in Burbank.

The future looked bright for the newest of Hollywood's moguls, until the stock market had the bad manners to crash in October 1929. The Great Depression that followed woke the Warners to the new realities of show business. Costs were slashed, and layoffs became the rule. By the end of 1932, Warner Bros. posted a loss of over $14 million, and the brothers feared their newly built empire would soon crumble.

This time, Rin Tin Tin would not be rushing to their rescue. Instead, Warner Bros. kept in business by cranking out cheap B-pictures along with a series of socially conscious gangster films and highbrow biopics. The famous *Looney Tunes* and *Merrie Melodies* cartoon characters debuted during these lean years from a laugh factory called "Termite Terrace," and the Warners furthered their foray into sound by bringing big Busby Berkeley–choreographed musicals to the screen. By the mid-1930s, Depression-era audiences had tired of musicals, and Warners' latest discovery, Errol Flynn, swooped in to swashbuckle their troubles away.

At the close of 1934, a fire engulfed the Burbank studio, destroying 20 years worth of irreplaceable WB, First National, and Vitagraph films. At around the same time, Benjamin and Pearl Warner both passed away. Jack and Harry had never been close, and their parents' deaths removed the last remaining buffer between them. This rift would eventually grow into an irreparable break between the brothers.

On the plus side, WB was able to scrape together a modest profit by 1935. Several of their films during these years were well received both commercially and critically, including 1937's *The Life of Emile Zola*, which won the studio its first Academy Award for Best Picture.

By the close of the decade, war clouds forming in Europe were largely being ignored in America. As a wake-up call to alert the country to the dangers of Nazism, the brothers produced *Confessions of a Nazi Spy* (1939). It certainly got the attention of the isolationist U.S. Congress, who called Harry to Washington to address charges of "creating hysteria." Within months of the film's release, Adolf Hitler's true intentions became clear to everyone when he invaded Poland and launched the century's second world war.

The immigrant Warners, who had witnessed totalitarianism first-hand in their native land, committed the full force of their media empire to helping win the war. During the conflict, they produced such patriotic films as *You're in the Army Now* (1941), *Yankee Doodle Dandy* (1942), *Now, Voyager* (1942), *This Is the Army* (1943), and 1943's *Casablanca*—one of Hollywood's crowning achievements.

Warner Bros. provided more than just films for the war effort. Nearly $20 million dollars in war bonds were purchased through the studio, and one quarter of WB's personnel, including Jack Warner and his son, served in the armed forces. The U.S. government rewarded their loyalty by christening a Liberty ship the SS *Benjamin Warner* in honor of the patriarch of the Warner family.

The first of the postwar years brought record profits to WB. The Warners' employees, who had sacrificed during the conflict, demanded a bigger share of those returns. The moguls would have none of it, and the workers responded by rioting outside the gates and walking off the job for a month. The studio chiefs were also having labor issues with some of their stars, like Olivia de Havilland, who successfully sued the Warners to break her contract. The ruling voided the contracts of several other actors as well, effectively ending the "studio system" under which Hollywood had operated since the early days.

In 1947, the good feelings the studio and the government had shared during the war were forgotten when the U.S. Justice Department found the major studios guilty of antitrust violations and forced them to sell their theater divisions. This ended a run of over 40 years—going back to the Cascade Theatre days—in which the company had been involved in film exhibition.

The government was not done. The brothers had produced a film during the war called *Mission To Moscow* (1943) as a favor to President Roosevelt. It was intended to encourage aid to Russia, who was then a World War II ally. With Russia the new enemy, and the country gripped by McCarthyism, Jack Warner was forced to appear before the House Un-American Activities Committee (HUAC) to testify about communism in Hollywood. Under pressure to name names, he implicated several WB employees, earning the lifelong emnity of many former friends in Hollywood.

By the dawn of the 1950s, all of the major studios faced their biggest threat to date from an upstart medium called television. Hollywood feared it would keep people in their living rooms instead of in the theaters, and most producers stuck their heads in the sand hoping the new technology would just go away. Conversely, WB chose to invest in the new medium, but this effort was halted by the government, fearing a new form of media monopoly.

After this rebuke, the brothers decided to circle the wagons with the rest of the Hollywood majors by refusing to show any of their films on television. WB experimented with 3-D and CinemaScope to lure audiences back to the theater, and for the first time, most Warner features were released in color. By 1955, the moguls threw in the towel and cautiously dipped their toes into the television waters with a program called *Warner Bros. Presents*. Eventually, the studio embraced the new medium, if not its stars, whom Jack Warner felt were "too independent." The studio launched several programs in the 1950s, mostly Westerns, such as *Cheyenne*, *Maverick*, *Bronco*, and *Colt 45*.

The Warner Bros. musical made a successful comeback during the 1950s, making Doris Day the studio's biggest star. Competition from television forced the studio to tackle more controversial subjects in their films, often redefining the role of the protagonist into that of an antihero. Marlon Brando made his motion picture breakthrough as Stanley Kowalski in 1951's *A Streetcar Named Desire* and Warner Bros. later took a chance on an up-and-coming 24-year-old Hoosier actor by the name of James Dean. He made three landmark films on the lot in the mid-1950s—*East of Eden* (1955), *Rebel Without a Cause* (1955), and *Giant* (1956)—before dying in a car crash.

By the middle of the decade, Jack and Harry Warner, the two brothers most dissimilar in age and temperament, were constantly bickering. The nadir of the struggle came when—in an episode straight out of one of their cartoons—an infuriated Harry chased Jack around the studio lot with a lead pipe. Something had to give.

The solution appeared to come from a 1956 buyout offer from a group of Boston investors. Jack convinced his two older brothers that it was time for the Warners to sell all of their shares and retire as a group. But in a secret deal, Jack bought all the shares back for himself after his brothers had signed the papers. He was now the largest single shareholder of the company and claimed the Warner Bros. presidency, which had always belonged to Harry.

From that day forward, neither corporate nor familial ties would ever bind the brothers again. Harry would die of a stroke brought on by the betrayal a short time later, and Albert, who lived for another 11 years, would never speak to Jack again.

In what some believe to be divine retribution, Jack was involved in a near-fatal car crash four days after Harry's funeral (which he skipped), but he lived to produce another day. Jack remained at the studio that he and his brothers founded for another decade before selling his remaining shares in 1966. He produced independently for a few more years until his death in 1978. Even after his passing, the family breach persisted. Jack was buried in the same East Los Angeles cemetery as the rest of the Warners, but not with them inside the family mausoleum.

The Warner Bros. story did not end with the passing of the brothers. Corporate mergers have made their company part of a much bigger media conglomerate, but the studio they purchased in 1928 in Burbank still exists. It is a vibrant living memorial to four young immigrant brothers who saw magic in sepia-toned images flickering across a bed sheet—and chose to create some magic of their own.

One

YOU AIN'T HEARD
NOTHIN' YET
1903–1928

THE BROTHERS WARNER, 1926. The four Warner brothers are, from left to right, Harry, Jack, Sam, and Albert. By 1926, they ranged in age from 33 (Jack) to 45 (Harry) and had been in the motion picture business together for over 20 years. They had recently purchased Vitagraph Studios and were busy making *Don Juan* (1926), their first film using the Vitaphone sound process.

THE BOSS OF THEM ALL
"MA" WARNER

HARRY M. WARNER

"DAD"
BENJAMIN WARNER

ALBERT WARNER

JACK L. WARNER

SAM L. WARNER

DAVID WARNER

WARNER FAMILY, 1921. In 1921, *Motion Picture News* ran this image showing pictures of the five Warner brothers (including David, who was not associated with the film business), along with father Benjamin and mother Pearl ("The boss of them all"). Benjamin and Pearl Wonskolaser were married in Krasnosielc (possibly Karsnashiltz), Poland, which was then part of the Russian Empire, in 1876. The couple had several children in Poland, including Harry (born Hirsch) in 1881, Albert (born Aaron) in 1884, and Sam (born Szmul) in 1887. Around 1890, Benjamin took a boat to America in search of a better life. Pearl and the children joined him a year later, and the family—now with the Americanized surname of Warner—settled in Baltimore. After a move to Canada, more children followed, including Jack (born Jacob or possibly Itzhak) in 1892. The family eventually moved to Youngstown, Ohio, where at various times they ran a cobbler shop, a butcher shop, a bicycle shop, and a grocery store. It was in Youngstown that the brothers began exhibiting films in rented halls.

CASCADE THEATRE, NEW CASTLE, PENNSYLVANIA, 1907. Harry Warner stands in front of the Cascade Theatre in 1907. The Warners opened their first permanent theater in this city because New Castle had no other theaters and was only a short ride from their home in Youngstown, Ohio. Figuring they could make more money renting films to other theaters than by showing them, they soon bought three trunks of films for $500 and opened a film distribution exchange in Pittsburgh. The Cascade was so successful that the Warners were able to sell their interest in 1909 for $40,000. They used the profits to expand their film distribution business by opening a second film exchange office in Norfolk, Virginia. Today a group of dedicated volunteers is working to restore the site to the way it looked during the Warners' time at the Cascade.

LKO Studios, 1920. After the collapse of their film exchange business, the Warners became producers on the East Coast. To fund the fledgling enterprise, they needed a legitimate hit. They got their wish in 1918 with the release of the World War I propaganda film *My Four Years in Germany*, based on a best-selling book by James W. Gerard. Somehow the Warners were able to convince Gerard that their studio was the best place to produce the film, even though their biggest success up to that point was a film made for the U.S. Army warning World War I soldiers about venereal disease, which, incidentally, was Jack Warner's only starring role on film. *My Four Years in Germany* cost $50,000 to produce and brought in 10 times that amount at the box office. The success of the film fueled their permanent move to the West Coast, where they made films with comedian Monty Banks (facing the camera at far right) here at the LKO Studios at the corner of Sunset Boulevard and El Centro Avenue in Hollywood.

Bostock Jungle and Film Company, 1915. One of the Warner's first Los Angeles studios was here at 1919 South Main Street near downtown. Built in 1887, and later known as Chutes Park, it was one of the West Coast's first amusement parks, with attractions, a zoo, theater, and baseball stadium—home to the Los Angeles Angels of the Pacific Coast League. The site later became David Horsley's Bostock Jungle and Film Company, which leased space to the brothers in 1919. Horsley, one of the film industry's true pioneers, built Hollywood's first studio in 1911, and later became one of the founders of Universal Pictures. He created the 5-acre studio at Chute's Park in July 1915 after acquiring the Bostock Zoo animals. When not filming, the animals were used in live shows for the public. The area was abandoned by 1925, and today is a parking lot.

ROMAYNE SUPERFILM COMPANY STUDIOS, 1919. The Romayne Superfilm Company Studios in Culver City was another early rental studio used by the Warners. The aerial photograph above shows the small lot with its partially covered studio stage, and the front entrance is featured in the image below. Romayne Studios was owned by producer Henry Y. Romayne. The lot was not ideal; Jack Warner called it "a dump." Warner Bros. produced a two-reel comedy at this studio starring Monty Banks. Romayne was not the first motion picture studio in Culver City; Thomas Ince built one there as early as 1915. MGM (today's Sony Pictures Studios), Selznick International Studios (where 1939's *Gone With the Wind* was made), and Hal Roach Studios soon became Culver City tenants. Romayne Studios was located on the northeast corner of Ince and Washington Boulevards.

WM. HORSLEY STUDIOS, 1920. The Warners became the newest "Poverty Row" producers in 1919 when they leased space from David Horsley's brother William, here at 6050 Sunset Boulevard. Next door, Horsley owned a film developing company called the Wm. Horsley Film Laboratories. The lab, now called the Hollywood Digital Laboratory, is the oldest existing company in Hollywood. The site later became the home of Columbia Pictures.

GORDON STREET STUDIO, 1919. The Warners moved from downtown Los Angeles to Culver City, and then to Hollywood in 1919. A short time later they moved again within Hollywood to the Gordon Street Studio on the south side of Sunset Boulevard and Gordon Street. In this photograph, actors gather at the gates hoping for a day's work in the "flickers."

WARNER BROS. SUNSET STUDIOS, 1922. In 1920, the Warners moved again, this time to a vacant lot in Hollywood fronted by Sunset Boulevard and bordered by Van Ness and Bronson Avenues, where they constructed this building (shown above, with a close-up of the entrance below). The 10-acre property was owned by a family named Beesemyer, who arranged for the brothers to pay for the $25,000 lot in installments. It was said that the payments were made in bags of gold coins. It did not look like much at the time, but 5842 Sunset Boulevard would soon be the place where Rin Tin Tin would first save the day, Bugs Bunny would nibble his first carrot, and where the voices of the stars on the silver screen would first be heard.

WARNER BROS. SUNSET STUDIOS, 1924. The Warners' new home in Hollywood changed dramatically in a short time. Only two years after occupying a single building on the grounds (see previous page), they expanded to fill nearly the entire 10-acre lot. The new multi-columned main studio building now bordered Sunset Boulevard, with Bronson Avenue on the building's right and Van Ness Avenue on the left side of the lot.

WARNER BROS. SUNSET STUDIOS (CLOSE-UP), 1924. The original studio building did not disappear but was built around, as seen in this close-up photograph. The studio site, which still exists today on Sunset Boulevard near the Hollywood Freeway (U.S. 101), was designated a cultural landmark in 1997 and placed on the National Register of Historic Places in 2002.

WARNER BROS. SUNSET STUDIOS (MAIN ENTRANCE), 1930. Behind these doors on Stage 3, *The Jazz Singer* (1927) was filmed. Nearby was the birthplace of the world-famous *Looney Tunes* and *Merrie Melodies* characters, brought to life in a rickety wooden building known as "Termite Terrace." The Warners continued to make films here until 1937, including James Cagney's *The Public Enemy* (1931).

JACK WARNER, 1922. J. L. Warner had a film career lasting over 70 years and was a studio head longer than anyone else in history. To some, he was a visionary, and to others, a tyrant. To all (except perhaps his brother Harry), he was the boss. For a half-century, he headed up production for Warner Bros. and ruled the studio as his own personal fiefdom.

THE WARNER BROTHERS, 1921. The Warner brothers, from left to right, are Sam, Harry, Jack, and Albert. The story of the Warner brothers is a tale of four impoverished immigrant boys who used their "advantages of disadvantage" to overcome the odds. By being Jewish, they faced prejudice, but countered it by succeeding in so-called "Jewish trades" that prepared them for the rough-and-tumble world of Hollywood. Being poor, they entered the workforce at young ages, keeping their formal educations to a minimum. This apparent drawback kept them unencumbered by set careers—any trade was as good as the next, as long as it paid the rent. This gave them the ability to act quickly on hunches and to swiftly embrace new ideas. But most importantly, their willingness to work together as a family combined with their good fortune to come of age alongside the motion picture industry quickly transformed the Warner brothers into Warner Bros.

THE "MORTGAGE LIFTER," 1924. German shepherd Rin Tin Tin is pictured here with writer Darryl F. Zanuck (left), Jack Warner (center), and and director Malcolm St. Clair, who examine a script for Find Your Man (1924). Trainer Lee Duncan discovered the shell-shocked puppy in France during World War I and named him Rin Tin Tin after a French puppet. In America, "Rinty" saved the Warners from bankruptcy by starring in several hit films in the 1920s. By 1926, he was earning $6,000 per week as the world's biggest box-office draw. Jack Warner nicknamed him the "Mortgage Lifter." A great deal of the Warners' success in the 1920s can be attributed to Darryl F. Zanuck. Zanuck began his long, illustrious career as a writer, producer, actor, director, and studio executive by writing the Rin Tin Tin scripts. An amazingly prolific screenwriter, Zanuck wrote over 40 scripts for the Warners during the 1920s. He became the head of production in 1931, but quit two years later to create what was to become 20th Century Fox Studios.

WARNER EAST HOLLYWOOD ANNEX, 1928. The Warners took their biggest gamble to date in 1925 when they purchased Vitagraph Studios. Vitagraph, one of the founding enterprises in the motion picture business, began in 1897 when Englishman J. Stuart Blackton, who worked as a reporter, was sent to interview inventor Thomas Edison. Edison talked Blackton into purchasing one of his projectors and some films. A short time later, Blackton founded the American Vitagraph Company with his business partner, Albert E. Smith, and went into direct competition with Edison. When Warner Bros. purchased Vitagraph, they obtained this studio (aerial view shown above and entrance below) at 4151 Prospect Avenue in the Los Feliz district near Hollywood. They sold it to ABC in 1948, and today it is known as the Prospect Studios, the home of many long-running soap operas and game shows.

KFWB Studios, 1925. In 1925, the studio entered into the Los Angeles radio market at roughly the same time as the Vitagraph merger. Contrary to popular legend, KFWB's call letters were not chosen ahead of time to stand for "Keep Filming, Warner Brothers," but just happened to be the next letters randomly issued by the government. KFWB's studios were for a time located on the Warner Bros. Sunset Studios in Hollywood. Later they were transferred to the Warner Hollywood Theatre at 6433 Hollywood Boulevard, as the broadcast signal interfered with the studio's new sound equipment. Warner Bros. sold the station in 1950, and it has since been owned by a string of media companies, most recently CBS Radio, which until recently operated KFWB as an all-news station. One of the first transmission towers still stands at the former site of the Sunset Studios in Hollywood. In this 1925 photograph, Jack (left) and Harry Warner broadcast from the KFWB studios on the lot.

DON JUAN, 1926. Jack Warner (left), John Barrymore (center), and Harry Warner pose together during the making of *Don Juan* (1926). *Don Juan* was Hollywood's first major sound film, but because none of the dialogue was spoken, it was not the first "talkie." Synchronized sound was only used in the film for background music and effects. Still, it paved the way for Hollywood's talkie breakthrough, *The Jazz Singer* (1927).

DON JUAN PUBLICITY, 1926. This publicity shot for the release of the synchronized recorded soundtrack for *Don Juan* (1926) was taken on August 7, 1926. From left to right are Jack Warner, Ernst Lubitsch, Will H. Hays, Harry Warner, and Sam Warner. Will Hays, Hollywood's chief censor, spoke on screen at the beginning of *Don Juan*. His was the first voice many theatergoers ever heard on film.

DON JUAN CAST AND PREMIERE, 1926. The cast of *Don Juan* (above), including the film's star John Barrymore (seated center), with Jack Warner standing behind, pose for this publicity shot at the Warner Bros. Sunset Studios in Hollywood. Barrymore, who was a gifted Shakespearian stage actor, was known to be difficult on film sets, feeling that the medium was beneath his talent. He behaved himself during the filming of *Don Juan*, knowing he was helping change the course of motion picture history. *Don Juan* debuted at the Warners' Theatre in Manhattan (left) to record-breaking audiences. Despite its successful premiere, it was unable to recoup its costs and left the Warners indebted. Unfazed, the Warners would produce *The Jazz Singer* the following year.

THE JAZZ SINGER, 1927. October 6, 1927, marked a tectonic shift in the history of the world's filmed entertainment. Before that day, silent films were king. After the world premiere of *The Jazz Singer* that evening at the Warners' Theatre in Manhattan, talkies would ascend to the Hollywood throne. While the idea of talking pictures was not a new one (some of the earliest films had soundtracks), silent films held many advantages over talkies. A silent film could be shot anywhere without the risk of a passing ambulance ruining a take, and a quick translation of dialogue cards could make it available to a foreign audience. Silents were also inexpensive to exhibit—an organist-for-hire was usually the only soundtrack that was ever needed. For these reasons, most believed silents would continue to rule Hollywood. That all changed on the night of the premiere. Sadly, none of the Warner brothers were on-hand to share the moment, because three of the brothers were rushing back to California to be with their dying brother Sam.

27

THE JAZZ SINGER, 1927. Al Jolson performs the song "Mammy" in blackface during the film *The Jazz Singer* (1927). Blackface performances were once a popular form of entertainment in vaudeville but are universally considered racist today for promoting negative stereotypes. Blackface performers are sometimes credited with bringing the music of African Americans *to* white audiences, yet more often vilified for keeping black musicians excluded from performing *for* white audiences.

WARNER HOLLYWOOD THEATRE, 1928. This theater, located at 6433 Hollywood Boulevard, was intended to host the West Coast premiere of *The Jazz Singer* (1927), but trouble with the new sound system postponed its opening. The film being shown on the day this photograph was taken was *Lights of New York* (1928), the first true "talkie." One of Hollywood's biggest blockbusters to date, it cost only $23,000 to produce, but sold over $1 million in tickets.

JACK WARNER AND AL JOLSON, 1927. By all accounts, Jack Warner (left) idolized Al Jolson, and it is easy to understand why. Both men were sons of Jewish immigrants from the Russian Empire who sought escape from poverty in show business. Both were singers—at least Jack fancied himself a singer. As a young man he tried to make a living singing in vaudeville under the name Leon Zuardo to very limited success. Jolson, on the other hand, was billed as the "world's greatest entertainer." That being said, the lead role in *The Jazz Singer* (1927) was originally offered to George Jessel, but after a disagreement over money, Jolson stepped into the role and made movie history. Jolson starred in a follow-up talkie for the Warners called *The Singing Fool* (1928), which featured "Sonny Boy," America's first million-selling song. The film was actually more successful than *The Jazz Singer*—so successful in fact that its box office take was not exceeded until *Snow White and the Seven Dwarfs* was released in 1937.

H.M. WARNER   ALBERT WARNER   S.L. WARNER   J.L. WARNER

SAM WARNER, 1918. Sam Warner (third from left) was the man most responsible for the emergence of sound films in the 1920s. A visionary (and possibly the most popular of the brothers), he believed sound was the next big thing to come to movies. Sam worked round-the-clock overseeing Warner's groundbreaking Vitaphone shorts and features. The stress ruined his health, and in what can only be called an untimely death, he passed away from a cerebral hemorrhage 24 hours before *The Jazz Singer's* 1927 premiere. There is disagreement over his age at death. Birth records were spotty in the Warner family (Jack Warner supposedly chose his own birth date), and most accounts list Sam as having been 39 or 40 when he died. His gravestone lists the years 1885–1927, which would have made him as old as 42.

NOAH'S ARK, 1928. The Warner East Hollywood Annex (formerly Vitagraph Studios) was often used for large productions like *Noah's Ark* (1928). Pictured here are the film's director, Michael Curtiz (left), and writer Darryl F. Zanuck. During the climactic flood scene, several extras suffered fractures, one had to have a leg amputated, and incredibly, three actually drowned. (John Wayne and Andy Devine were two of the lucky extras who escaped injury that day.)

WARNER VITAGRAPH STUDIOS BROOKLYN, 1924. Warners' purchase of Vitagraph gave them an additional studio in Brooklyn, New York, at East Fifteenth Street and Locust Avenue in the Midwood section. Some of the very first "talkie" short subjects were made here under the direction of Sam Warner. It was sold in 1939, and now the Shulamith School for Girls occupies the site. The smokestack with the word "Vitagraph" seen at the upper-right corner of the photograph still stands.

WARNER BROS. SUNSET STUDIOS, 1926. Three studio division heads pose with Jack Warner next to the KFWB radio tower in front of the Warner Bros. Sunset Studios. From left to right are Darryl F. Zanuck, who functioned as production manager; a nattily-dressed Jack Warner; Bryan Foy, who headed up the studio's B-picture division; and Leon Schlesinger, who was in charge of the cartoon unit. Bryan Foy began his career in vaudeville as part of "Eddie Foy and His Seven Little Foys." Leon Schlesinger headed up Leon Schlesinger Productions, which later became Warner Bros. Cartoons. He was the man who brought together the legendary talents of Friz Freleng, Tex Avery, Chuck Jones, Bob Clampett, Mel Blanc, and Carl Stalling—the men who gave birth to Daffy Duck, Porky Pig, and Bugs Bunny. These world-beloved characters came to life from a small building on the lot near the corner of Van Ness and Fernwood Avenues called "Termite Terrace."

WARNER BROS. SUNSET STUDIOS, 1927. The look of the lot changed a great deal during the 1920s (aerial view above and main building below). What was once vacant land grew into a complete film factory. These photographs were taken the year *The Jazz Singer* was released. The film proved so successful that the Warners found themselves rich for the first time in their careers. Always looking to expand, they saw low-hanging fruit waiting to be plucked in the form of financially troubled First National Pictures. They moved most operations to First National's new studio in Burbank after the purchase and sold the Hollywood property in 1937. For years it was a bowling alley, but eventually reopened as a studio, passing through several owners, including Gene Autry and Paramount Pictures. Today it is known as Sunset Bronson Studios.

BURBANK RANCH HOUSE, 1908. Dr. David Burbank was a dentist from New Hampshire who moved to the area now known as Burbank in 1866. The following year, he purchased 9,200 acres across two land grants from the Spanish and Mexican eras, built this ranch house, and began raising sheep and wheat. In 1926, First National bought a 78-acre parcel from rancher Stephen A. Martin to build their new studio. This included a 40-acre hog farm and the original Dr. Burbank house, seen here in 1908. Much has changed on the old homestead in just over a century. Today the Burbank ranch site is not only the home of Warner Bros., but it is the heart of Burbank's Media District. Several radio stations, record companies, and television studios—including the home of NBC's *Tonight Show with Jay Leno*—are now located here.

FIRST NATIONAL STUDIOS BURBANK, 1926. Here are two of the earliest views of the construction of the First National Studios in Burbank in 1926. Above, the aerial photograph facing eastward shows that much of the surrounding area was still uninhabited. Today a freeway (California 134) cuts through the fields lying parallel to the left side of the photograph, and the Walt Disney Studios occupy the area at the upper left-hand corner. The photograph below was taken on May 26, 1926, and shows two partially constructed buildings along with the completed water tower. The tower stood at this spot for many years before being moved to its current position closer to the north side of the lot.

FIRST NATIONAL STUDIOS BURBANK, MAY AND JUNE 1926. Thomas L. Tally founded First National Pictures in 1917 by merging 26 theater chains together to take on Paramount Pictures, which was then the biggest exhibitor in the business. During the next two years, First National signed million-dollar distribution deals with Mary Pickford and Charlie Chaplin. By 1924, First National began producing films, and two years later commenced building a new 62-acre lot in Burbank, 6 miles north of Hollywood. The above image from April 1926 shows a stage being built. The photograph below shows a stage nearing completion.

FIRST NATIONAL STUDIOS BURBANK, 1926. The June 1926 photograph above shows a view of the studio's construction looking southwest. The photograph below is of the facade and colonnade of the new wardrobe department. The land to the west of the studio has figured prominently in the history of California as the site of a military skirmish that helped oust the Spanish governor of California, who was replaced by Mexican leader Pio Pico. It is said that locals living near the studio dug up evidence of the battle years later in the form of cannon balls.

FIRST NATIONAL STUDIOS BURBANK, AUGUST 1926. The image above shows one of the dressing room buildings that First National constructed. The photograph below features an administration building that eventually housed studio office workers. Not everyone was happy with First National's move into production. Both the Independent Producers Association and the Motion Picture Theatre Owners of America claimed that First National (along with MGM and Paramount Pictures) was part of a film trust seeking to dominate distribution, exhibition, and now production as well.

FIRST NATIONAL STUDIOS BURBANK, 1927. These two images show the freshly completed original Stage 5 (above) and the new east administration building, along with dressing room buildings 1 and 2 (below). The Warners finished *The Jazz Singer* (1927) in Hollywood around the time of these photographs. *The Jazz Singer* whetted the audiences' appetites for talkies, and the Warners followed up this blockbuster with several more, including *Lights of New York* (1928), which has the distinction of being the first "all talkie" feature; Al Jolson's *The Singing Fool* (1928), the biggest box office talkie until 1937; and *The Terror* (1928), the first sound suspense film. These successes doomed the silent era and secured gargantuan loans for the Warners, which they used to purchase First National.

First National Studios Burbank, 1927 and 1928. These two photographs show construction of First National soundstages (the new preferred term for stages) between February 1927 and September 1928. It is unlikely that First National got much use out of them for their own productions, because September 1928 was the same month that Warner Bros. bought a majority interest in the studio. The purchase gave the Warners control of the Burbank lot and First National's chain of theaters. For tax reasons, the companies functioned as distinct entities until 1936. Until then, "First National Pictures" still appeared on the screen at the beginning of all of their films.

FIRST NATIONAL STUDIOS BURBANK, 1926 AND 1929. The aerial shot above shows the studio on June 23, 1926, while construction on the lot was still taking place. The photograph below is from 1929, after the lot was acquired by Warner Bros. In this rare photograph looking north from the side of Mount Lee, the completed studio's soundstages, administration buildings, and backlot in the lower right corner are clearly visible. The spaciousness of the San Fernando Valley is striking, particularly to anyone who has spent time stuck in traffic on its freeways. Part of the reason for today's crowded conditions in "the Valley" can be attributed to the success of Warner Bros. Other studios soon followed, most notably Walt Disney, which made Burbank its home in 1940.

First National Studios Burbank, 1929. After nearly a decade in Hollywood, the Warners found a new home in Burbank. The First National name would linger on for another three decades as most WB films were credited as "Warner Bros.-First National Picture" releases. After that, the company's name would be all but forgotten. Here is a look at the main administration building in 1929 after a rare snowfall.

Harry Warner, 1926. Harry Warner secured the loans that kept the company afloat in bad times and got them out of "Poverty Row" in good ones. Tired of bickering with Jack, Harry left Burbank in 1929 to manage things from New York alongside Albert, who ran the theater division from there. Afterwards, he seemed a happier man with an entire continent serving as a buffer between himself and his youngest brother.

Two

BURBANK

THE ROARING THIRTIES

MOUNT WARNER, 1930. The mountain overlooking the Warner Bros. Studios in Burbank was renamed "Mount Warner" in 1930 as a tribute to Sam Warner (who passed away in 1927) and his three brothers, for their contribution to talking pictures. From left to right are Jack Warner; Jack Warner Jr., who prepares to christen the monument with a champagne bottle; and Benjamin Warner, the father of the Warner brothers.

WARNER FIRST NATIONAL STUDIOS (AERIAL PHOTOGRAPH), 1932. After scraping by for nearly two decades as a second-tier producer, distributor, and exhibitor, Warner Bros. had finally made the big time with their purchase of First National. But Warner Bros.'s arrival on the Hollywood scene nearly proved to be short-lived as the Great Depression, which began soon after, almost forced the new enterprise out of business.

WARNER FIRST NATIONAL STUDIOS (AERIAL CLOSE-UP PHOTOGRAPH), 1932. Mines Field opened in 1928. Its name was changed to Los Angeles Airport in 1941, and later to Los Angeles International Airport (LAX) in 1949. The arrow with the words "L.A. MINES 14M" atop the soundstage aided pilots in navigating to the new field, 14 miles to the southwest.

CENTRAL POWER STATION, 1939. The Central Power Station, seen here in 1939, sits next to the Warner Bros. Fire Department and water tower, near the center of the original layout of the studio. The original power plant had a 12-billion candle power capacity. Today the plant and fire department remain, but the water tower has been moved to the northwest.

JACK WARNER, 1936. At times tyrannical and narcissistic, at others charming and charitable, Jack Warner was the *real* power source at the studio. Without his favorite brother Sam to mediate, he was constantly at odds with Harry, who called the financial shots for the company from New York. Despite his lack of formal education, Jack had an innate sense of story and produced hundreds of successful films.

PERSONNEL ENTRANCE, 1937. At one time, this guard shack stood between the main administration building and the part of the studio containing the soundstages, production and services buildings, and backlots. The guards are no longer here, but the arch remains as part of Building 9.

VITAPHONE WRITERS, 1930. Pictured are the men who wrote the Vitaphone shorts in the 1930s. Among the group are studio executive Darryl F. Zanuck (standing, first row) and writers Gordon Rigby, J. Grubb Alexander, Joseph Jackson, Rex Taylor, Howard Smith, Perry Verkroff, Walter Anthony, Lucian Hubbard, F. Hugh Herbert, Charles Kenyon, Harvey Thew, and Cyril Hume. This photograph was taken at the Warner Bros. Sunset Studios in Hollywood.

ANTHONY ADVERSE EXTERIOR, 1935. *Anthony Adverse* (1936) was a period costume drama directed by Mervyn LeRoy that starred Fredric March and Olivia de Havilland. The dock scenes were filmed on Bonnyfeather Street (pictured above and below), which was named after a family of characters in the film. Bonnyfeather Street was located on the Warner Bros. backlot in an area that was once near what is now Hennesy Street. The film was nominated for seven Academy Awards that year, including best picture, which it lost to MGM's *The Great Ziegfield*. *Anthony Adverse* reportedly required 131 sets, including a 12-acre African compound on the backlot that was one of the largest sets ever constructed.

ANTHONY ADVERSE INTERIOR, 1935. Mervyn LeRoy (front left) directs a dance scene for *Anthony Adverse* (1936). LeRoy's directorial career lasted from 1927 until 1968, when he helped out on *The Green Berets* (1968). Along the way, Warner's "Boy Wonder" helmed many of the studio's biggest hits and married Harry Warner's daughter. As head of production for MGM, he produced *The Wizard of Oz* (1939).

DR. SOCRATES, 1935. This photograph comes from the set of the Paul Muni crime thriller *Dr. Socrates* (1935). This scene was shot in an area of the backlot that eventually was rebuilt as Midvale Street in 1939, and later renamed Midwest Street. It was here that the band (including coauthor Marc Wanamaker) marched in *The Music Man* (1962).

STAGE 7 EXTERIOR, 1936. Warner Bros. literally "raised the roof" on Stage 7 (now Stage 16, pictured above and below) in 1936 for the romantic comedy *Cain and Mabel* (1936). The film, which starred Marion Davies and Clark Gable, was billed as "Warner Bros.'s biggest musical extravaganza in three years" with "500 of the world's most beautiful girls." In spite of these claims, the film was a disappointment, and Davies would make only one more movie the following year before retiring from Hollywood. To film the intricate dance numbers, workers painstakingly jacked up the roof an additional 35 feet—a stunning achievement that was used as a selling point in the film's trailer.

STAGE 7 INTERIOR, 1936. Workers stand inside the gargantuan soundstage while it is being raised. Stage 7 (now 16) has a sunken floor that can be filled with 2-million gallons of water. This huge set has been the site of many landmark films, including *Yankee Doodle Dandy* (1942), *Ghostbusters* (1984), *Jurassic Park* (1993), and parts of the *Indiana Jones* franchise.

CLARK GABLE, MARION DAVIES, AND LLOYD BACON, 1936. Even the stars came out to see the soundstage being raised. *Cain and Mabel's* director Lloyd Bacon (far right), points out work being done on the stage to the film's stars, Clark Gable (left) and Marion Davies. Davies's benefactor, William Randolph Hearst, reportedly paid the $100,000 it took to raise the stage out of his own pocket.

STAGE 7, 1936. This photograph shows the Mill building on the left, Stage 17 (Stage 21 today) on the right, and a fully raised Stage 7 (Stage 16 today) in the center. Fans of rock 'n' roll may remember this view from the "burning deal" photograph from Pink Floyd's 1975 *Wish You Were Here* album, which was taken from the center of the street near Stage 7.

WB COMMISSARY, 1937. Throughout the years, the lot has maintained a barbershop, shoe-shine parlor, post office, health club, several stores, and a top-notch commissary. The best place for star sightings has always been here, as well as next door in the exclusive dining room, once known as the "Green Room."

WB COMMISSARY, 1931. Adm. Richard Byrd (center) visits the Warner Bros. Studios and lunches in the company of George Arliss (left) and John Barrymore, who arrived in full makeup from the set of *Svengali* (1931). Barrymore passed away in 1942, and his drinking buddy, Errol Flynn, later portrayed him in 1958's *Too Much, Too Soon*.

WB COMMISSARY, 1937. Joan Blondell (left), Dick Powell (center), and Edward Everett Horton dine together in this photograph from 1937. Blondell and Powell costarred that year in *Gold Diggers of 1937*, the fourth in Warner's series of "Gold Digger" films. Horton was a longtime character actor who is often remembered as the narrator for Jay Ward's *Fractured Fairy Tales* segment of *The Rocky and Bullwinkle Show* (1959–1964).

WB COMMISSARY, 1938. Actors Errol Flynn (left), Margaret Lindsay (center), and Melville Cooper dine during the making of *The Adventures of Robin Hood* (1938). *Robin Hood* was one of the first three-strip Technicolor films to be made by Warner Bros., and its large budget was a departure from the studio's usual frugality. James Cagney was originally slated for Flynn's part but lost the role due to a contract dispute.

CARL JULES WEYL, 1938. Warner art director Carl Jules Weyl is here at work on set designs for *The Adventures of Robin Hood* (1938). Weyl was born in Stuttgart, Germany, and won an Oscar for best art direction for his work on the film. He was also nominated in the same category for the controversial 1944 Warner Bros. film *Mission to Moscow.*

THE ADVENTURES OF ROBIN HOOD, 1938. Errol Flynn and Olivia de Havilland are seen here being directed by William Keighley during the making of *The Adventures of Robin Hood* (1938). Errol Flynn's arrival on the lot during the middle part of the decade ushered in a new kind of Warner Bros. film: the swashbuckler.

WB STUDIOS, 1937. Here is a photograph of Avenue D, looking south from the area that is today the Eastwood Scoring Stage. The commissary is the building to the far left, and the fourth soundstage on the right is Stage 7 (today Stage 16), which had been raised to its current height of 98 feet the previous year.

WONDER BAR, 1934. Al Jolson (left) is greeted by studio executive William Koenig (center) and Dick Powell (right), while Ruby Keeler and an unidentified man look on from inside the car. *Wonder Bar* (1934) was another huge Jolson moneymaker for Warner Bros. It featured several Busby Berkeley–directed dance sequences, and almost did not fly with the censors, who considered it risqué. Jolson was the first superstar of the talkies, but within a few years, the novelty of his films wore off. He left Hollywood briefly to perform the stage musical *Wonder Bar*, then returned to WB to re-create the role on film. Jolson's last musical on the lot was *The Singing Kid* (1936), in which he parodies himself. Jolson and Keeler, the star of Warner's *42nd Street* (1933), were married from 1928 through 1939.

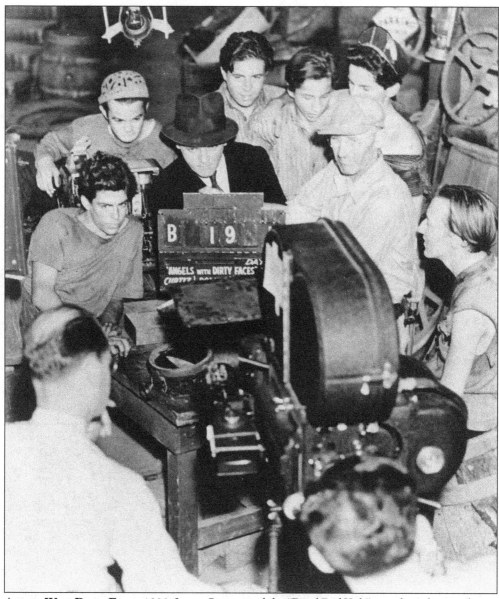

ANGELS WITH DIRTY FACES, 1938. James Cagney and the "Dead End Kids" pose for a close-up during the making of *Angels with Dirty Faces* (1938). The Dead End Kids got their name on Broadway in the play *Dead End*. They were brought to Hollywood to revive their roles in a film version for United Artists and bounced around Hollywood at several studios, performing at various times as the "East Side Kids," "Little Tough Guys," and "Bowery Boys." As possible reflections of their on-screen personas they earned the reputation of troublemakers on the lot as well. From left to right facing the camera are Billy Halop, Leo Gorcey, James Cagney, Bernard Punsley, Gabriell Dell, Bobby Jordan, an unidentified man, and Huntz Hall.

THE CABIN IN THE COTTON, 1932. Bette Davis arrived on the Warner Bros. lot in 1932 after a brief, unsuccessful stint at Universal. *The Cabin in the Cotton* (1932) was one of her first of 54 roles for Warner Bros. The photograph above shows filming of the exterior of the film's Norwood home on the studio lot. In the picture at right, Davis clowns behind a klieg light during the production. Davis was a ubiquitous part of the Warner landscape for years and is nearly as well known for her famous feuds with Jack Warner as she is for her legendary performances. Even in death, the "Fifth Warner Brother" is said to keep an eye on things at the studio from her grave, which faces the lot from a short distance away.

BROWNSTONE STREET, 1938. Brownstone Street is pictured here being used as a backdrop for a period picture. Brownstone Street was one of the first backlot sets to be constructed at the studio and has been seen in literally thousands of films, television shows, commercials, and music videos over the years. The street still exists, although greatly altered.

JUNE TRAVIS, 1935. The country was mired in the Great Depression in 1935, but beauty could still be found at Warner Bros. Studios. In this staged photograph, June Travis, a Warners' B-picture starlet, gets a speeding ticket from studio policeman Duke le Due. Travis was the daughter of a Chicago baseball executive and was discovered by a talent scout at a White Sox exhibition game in Miami.

GEORGE RAFT AND EDWARD G. ROBINSON, 1936. Warner Bros. was known as the "Gangster Studio" in the 1930s, and pictured here are two of the chief wise guys. George Raft (left) came from New York's tough Hell's Kitchen neighborhood and almost lived a life of crime instead of portraying one on the screen. Edward G. Robinson was a cultured art aficionado who hated the sound of gunfire.

DANGEROUS, **1935.** Bette Davis and Franchot Tone rehearse their dialogue for *Dangerous* (1935) with director Alfred E. Green. Davis would win the Oscar for best actress for this role. Until the end of her life she would contend that the statuette was a consolation prize from the Academy for not being nominated the previous year for her portrayal of Mildred Rogers in *Of Human Bondage* (1934).

SOUNDSTAGE, 1936. This photograph, taken outside of a studio soundstage in 1936, mixes several Warner Bros. stars in with a group of visiting VIPs. Included in the picture are Humphrey Bogart, Billy Mauch, Bette Davis, Bobby Mauch, Frank McHugh, Warner Bros. president Harry Warner, and studio executive Hal B. Wallis.

GOLD DIGGERS OF 1933. Musical director Leo Forbstein oversees the scoring for *Gold Diggers of 1933* (1933) while director Mervyn LeRoy and actress Ginger Rogers look on. *Gold Diggers* is a lavish Busby Berkeley–choreographed fantasy-musical with a gritty side, referencing the Great Depression, which was bottoming out at the time. A highlight of the film is Rogers's version of "We're In the Money" sung in Pig Latin.

BARHAM GATE, 1933 AND 1974. The Barham Gate (now Gate 2) is still the main employee entrance onto the lot. The old fashioned traffic signs were replaced ages ago, and the water tower in the photograph above (which bears the "First National Pictures" and "Vitaphone" logos) is no longer at this site. It was moved in 1937 a few hundred feet northwest to a location north of the Warner Bros. Museum. The photograph below is another famous view of the Barham Gate taken just over four decades later for the conclusion of Mel Brook's Western parody, *Blazing Saddles* (1974).

EVER SINCE EVE, 1937. In this scene from *Ever Since Eve* (1937), Marion Davies plays Marge Winton, a beautiful secretary who dresses down to thwart the advances of her employers. This was Davies's last film, closing out a 20-year Hollywood career. Her lover, publisher William Randolph Hearst, tried to make her a major star and created Cosmopolitan Pictures largely for that purpose, but she never quite made it with the public. Cosmopolitan was for a time partnered with MGM in Culver City, but after a falling out with Irving Thalberg, Hearst moved it to Burbank, beginning a two-year partnership with Warner Bros. This collaboration left a lasting legacy on the lot in Stage 7 (now Stage 16) when the roof was jacked up 35 additional feet for the Davies film *Cain and Mabel* (1935). Cosmopolitan Pictures folded in the 1930s, but its name lives on in the periodical published by Hearst Magazines: *Cosmopolitan*.

COLLEEN, 1936. Dance director Bobby Connelly (far right, under the camera platform) directs a complicated dance number for the film *Colleen* (1936), starring Dick Powell and Ruby Keeler. *Colleen* was a late-1930s musical, which was made at a time when audiences were turning to action pictures. Warner Bros. was increasingly feeding that hunger with swashbucklers, while phasing out musicals.

WARNER BROS. COLLAGE, MID-1930S. Jack (left), Harry (center), and Albert Warner (Sam Warner died in 1927) are pictured surrounded by images from 1930s WB films. Note the "Warner Brothers/First National & Cosmopolitan Productions" lettering on the soundstages. This was one of the last times the word "Brothers" would be written out in its entirety as part of the name. The partnership with Hearst's Cosmopolitan Productions ended in 1937.

CAPTAIN BLOOD, 1935. Olivia de Havilland (left) is fitted for a costume by employees of Warner's wardrobe department during the making of *Captain Blood (1935)*. In the photograph below, Guy Kibbee (left), Errol Flynn (center), and Ross Alexander turn a water wheel, while Michael Curtiz (foreground center) directs the scene. Flynn stormed out of the gate in *Captain Blood* and became a sensation with the fans (if not with Jack Warner, who would spar with him for years). De Havilland was discovered a short time earlier, and while best known today for playing Melanie Wilkes in *Gone With the Wind* (1939), she left a more far-reaching legacy in Hollywood in the 1940s by successfully suing Warner Bros. in a contract dispute. The ruling, still known as the "De Havilland Law," greatly reduced the power of the studios vis-à-vis actors.

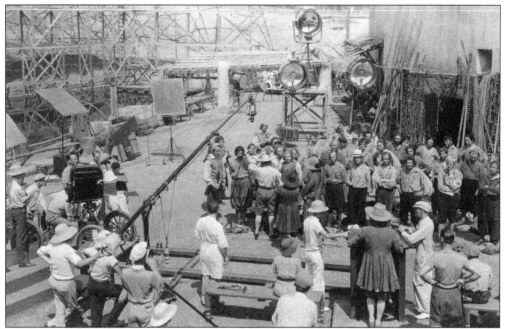

CAPTAIN BLOOD, 1935. One of the films made at the Warner East Hollywood Annex lot (formerly Vitagraph Studios) was *Captain Blood* (1935), which made Australian newcomer Errol Flynn a star. In the photograph above, Flynn is in the second row of prisoners, just to the right of center. The photograph below is of a wharf scene shot in the backlot tank. Vitagraph Studios was built in 1915, and two of the first stars to work there were Wallace Reid and Stan Laurel. In 1925, Warner Bros. bought the company and added this studio to its roster. It was sold in 1948 and is now known as Prospect Studios. This was the longtime home of *The Dating Game, Let's Make a Deal, American Bandstand, General Hospital,* and *The Lawrence Welk Show.* More recently, *Grey's Anatomy* has been filmed here.

THE WORLD CHANGES, 1933. One of the hundreds of Westerns made on the Burbank lot was Paul Muni's *The World Changes* (1933), directed by Mervyn LeRoy. Muni, like the Warners, was a Polish-Jewish immigrant. A brilliant stage actor, he hit pay dirt earlier for WB in *I Am a Fugitive From a Chain Gang* (1932), a film which helped reform prison practices in the South.

WB BACKLOT, 1930. A group of "soldiers" gathers for battle on a studio backlot. This area, near what is today the Hennesy Street backlot, was largely destroyed in a fire—a constant threat at the studio, which is why a fully equipped fire department resides on the lot. In spite of this precaution, major fires in the 1930s and 1950s destroyed large sections of the studio.

ENGLISH STREET, 1938. Once upon a time, in a magical place called Burbank, a person could enter the gates, meander around a couple of soundstages, and arrive in Merry Old England. Notice the man holding a slate with the words "MAT SHOT" in the middle of the street. This indicates that the scene would employ a matte painting to fill in areas outside the action.

WB STUDIOS, 1935. Gathered here at the studio are, from left to right, Paul Muni, Al Jolson, director Mervyn LeRoy, and production manager Hal B. Wallis. Wallis joined Warner Bros. in the 1920s and soon became the head of the publicity department and later production chief. He eventually formed his own company and made a fortune producing several Elvis Presley movies.

BETTE DAVIS, 1932. Bette Davis is fitted for a gown by employees of the studio's wardrobe department. Davis made 54 films for the studio and won two Oscars along the way. She had several legendary fights with Jack Warner, culminating in an unsuccessful bid to break her contract in an English courtroom.

KFWB 10TH ANNIVERSARY, 1935. Al Jolson oversees the festivities at KFWB's 10th anniversary party in 1935. The station was originally located on WB's Hollywood lot but soon moved to the Warner Theater in Hollywood, as the signal interfered with the studio's sound films. Among the group are Benny Rubin, Bette Davis, Warren William, George Brent, Lyle Talbot, Harry Warren, Bill Ray, Red Walker, Dick Powell, and Wini Shaw.

WB STUDIO CHIEFS, 1936. Hal B. Wallis (left), Jack Warner (center), and Harry Warner are pictured here in 1936. Wallis became production head after Darryl F. Zanuck's departure and oversaw several legendary Warner films, including *Captain Blood* (1935), *The Adventures of Robin Hood* (1938), and *Casablanca* (1942). He would later leave the studio after fighting with Jack Warner over money and the ownership of *Casablanca's* best picture Oscar.

MPTOA DELEGATION, 1934. Some of Warner's top executives and stars meet with delegates to the National Convention of the Motion Picture Theatre Owners Association of America during their visit to the WB Studios in 1934. Among the group are Jack Warner, Lyle Talbot, Phil Regan, Ruby Keeler, Kay Francis, Dick Powell, Ricardo Cortez, Dolores Del Rio, Mary Astor, Paul Muni, and Jean Muir.

MILL BUILDING, 1938. The Mill building was constructed in 1937 to consolidate several studio functions under one roof. This was done in response to a devastating fire in late 1934, which destroyed most of the backlot areas and individual service offices. The Mill is one of the largest buildings on any studio lot and today houses the paint department, sign shop, carpentry shop, photography lab, labor department, and a store. Since everything on a studio lot eventually makes it on film, it is not surprising to learn that the Mill has been seen in the movies. In *Giant* (1956), when James Dean and Rock Hudson fight in a wine cellar, it was actually right here in the Mill's basement. These photographs were taken from the building's south side, which fronts the Los Angeles River.

WESTERN TOWN, 1939. At various times, the studio's Western Town consisted of three parts: Western Street, Mexican Street, and Laramie Street. On these mean streets, thousands of gunmen died with their boots on after shootouts with men in white hats. Laramie Street, the last remaining part of Western Town, was bulldozed in 2004 and replaced with a row of production offices.

VIEW OF COMMISSARY, 1939. This view looks south towards the commissary building in the center, with soundstages off in the distance. This photograph was taken from the area between today's buildings 71 (left) and 6 (right). Building 71 now contains Screening Rooms 4 and 5, while Building 6 is now called the Eastwood Scoring Stage.

STUDIO GATES, 1936. In the photograph above, a guard patrols the entry gate from Warner Boulevard, which was used primarily by VIPs and known for a time as the "Marion Davies Gate." The photograph below shows the revamped Barham Gate (Gate 2 today), with the property department to the south and the construction of Stage 1 to the north. Originally, the studio was set far inland from Barham Boulevard (now Olive Avenue), with parking lots between the street and the soundstages. This changed in the mid-1930s with the construction of the new buildings, which now border the street. The property department still resides in this building, as well as the drapery and upholstery department and the security office. *Ellen* is currently housed in Stage 1, along with two adjoining stages.

ADMINISTRATION BUILDING, 1930 AND 1939. A decade separates these two views of the main administration building. The photograph above from 1930 shows the exterior a few months after the Warners had moved most of their operations here from Hollywood. At that time, this view was visible to motorists passing along Warner Boulevard. Since then, this area of Warner Boulevard has been incorporated into the studio grounds, and the lot now extends all the way north to Olive Avenue. Today the walls along Olive Avenue largely block this building from view. Below is a view of the main entrance to the building from 1939.

NEAR BARHAM GATE, 1936. In the photograph above, guards stand in front of the Barham Gate (now Gate 2). This setting looks largely the same today. The photograph below shows the view across the street from the guard shack. It is largely unaltered today as well, except due to the unending cycle of building renumberings that is the Warner Bros. way, Stage 22 is today Stage 1, and Stage 4 is now Stage 5. Over the years, Stage 22/1 has been used for *Yankee Doodle Dandy* (1942), *The Treasure of the Sierra Madre* (1948), *Rebel Without a Cause* (1955), and *Cool Hand Luke* (1967). It was also the television home of *ER* (1994–2009) for many years and is currently one of the stages where *Ellen* is taped.

PROPERTY DEPARTMENT, 1937. The property department building on Avenue A was built on space that had previously been parking lots in the mid-1930s. This brought the studio perimeter all the way out to Barham Boulevard (now Olive Avenue). Inside these walls can be found furniture and other large props that are used for Warner Bros. productions. One of the Vitaphone mobile sound recording trucks sits parked at the loading dock in the photograph above. The bottom view shows the same building looking north. In the distance is today's Studio 1, which is shown here under construction.

PROPERTY DEPARTMENT, 1938. The property department does not just house props—it sometimes creates them as well. In this photograph from 1938, prop shop workers construct hand props and breakaway chairs. The property department still occupies the same building on Avenue A, fronting Olive Avenue. The building is also home to the security office.

DRESSING ROOMS, 1937. In 1937, this building at the corner of First Street and B Avenue was the site of dressing rooms, as well as a shoe-shine parlor. Today it is the home of support offices, including the studio's credit union. Most current productions use mobile trailers called "Star Waggons" for dressing rooms. The extra "g" in the name pays homage to the company's owner, actor Lyle Waggoner.

WARNER BROS. SCHOOL, 1939 AND 1970S. The state of California mandates that every child actor must complete a specified number of hours of schoolwork while employed on a production. The building above once served as the Warner Bros. school. In the 1970s, it was used as an office for Burt Reynolds and is currently the studio's fitness center. In more recent years, portable trailers usually serve as schoolrooms. In the 1970s photograph below, taken outside of Stage 26, the trailer to the right was the school for the young actors from *The Waltons* (1972–1981). Notice the words "Walton's Mountain School" on the side. Incidentally, the mountain seen in the distance through the fog sometimes doubled as Walton's Mountain, and the Walton family home was located on the studio's jungle set.

THE SEA HAWK STAGE, 1939. Construction began on a new massive sound and water stage in 1939 for Errol Flynn's *The Sea Hawk* (1940). Stage 21 would eventually become the largest stage in the world. It occupied an area just to the west of today's Hennesy Street. The stage would remain at this site until 1952.

TRANSPORTATION DEPARTMENT, 1939. The studio's transportation department is responsible for the maintenance of the lot's fleet of 450 vehicles and for shuttling actors to and from off-site locations. This photograph from 1939 shows the department office (left) at the corner of Avenue A and Sixth Street, with gas pumps in the center and the Mill building in the distance.

STAGE 12, 1930. Two studio workers take a break outside of Stage 12 in 1930. Stage 12, at the corner of Second Street and Avenue E, was known for many years as Stage 24 but is now called "The Friends Stage" as a tribute to the long-running television comedy that was filmed here from 1994 to 2004.

ART DEPARTMENT, 1938. Here is a peek inside the Warner Bros. art department, with models and blueprints used to create sets for studio productions. On the table rests what appear to be scale models of a castle (complete with drawbridge), a building front, and a city street scene. The art department is located at the north end of Stage 15.

BLACK LEGION, 1937. This is a shot from an exterior scene for *Black Legion* (1937). The film starred Humphrey Bogart, who played a factory worker who joins a hate group that employs intimidation against foreign-born workers. In this photograph, the Mill building on the right doubles as a factory. Bogart made this film in the year between *The Petrified Forest* (1936) and *Angels with Dirty Faces* (1938).

ADMINISTRATION BUILDING LOBBY, 1939. One can still almost hear the echoes from the knocking knees of nervous Hollywood hopefuls who once waited in this room. This streamline art moderne-style lobby, with its glass wall, stainless steel door, and globe lamp overhead, gave many their first look inside Warner Bros. Notice the marquee on the left announcing the 1939 releases of *The Kid From Kokomo* and *Sons of Liberty*.

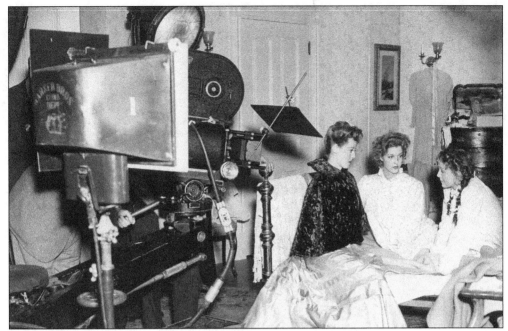

THE SISTERS, 1938. Two of Warner's biggest stars, Bette Davis and Errol Flynn, were paired together as husband and wife in *The Sisters* (1938), a romantic drama set in the early years of the 20th century. The film was originally slated to be billed as "Errol Flynn in *The Sisters*," but Davis demanded equal billing. From left to right are Bette Davis, Anita Louise, and Jane Bryan.

STAGE 5, 1939. Stage 5 (now Stage 15) on Third Street was part of the second phase of buildings constructed on the lot. The art department offices have been located here for many years and were constructed on the north side of the building to take advantage of the sunlight. Third Street is the main east-west artery of the studio coming in from the Barham Gate (now Gate 2).

BARHAM BOULEVARD, 1930s. Pictured above is a row of cars along Barham Boulevard (now Olive Avenue) after a rare snowfall in 1932. At the time, the studio was still offset from Barham Boulevard, and parking lots occupied the area between the soundstages (visible here) and the street. In the background is a view of Mount Lee to the south of the studio. This mountain has appeared in countless productions and was often used as "Walton's Mountain" for the television series *The Waltons* (1972–1981). Today it is crowned with transmission towers, and its opposite side displays the world famous Hollywood sign. (Ida Lupino and Errol Flynn once climbed this hill and threw rocks down at the studio.) Years later, the property department building (below) filled in the former parking lots.

WARNER FIRST NATIONAL STUDIOS, 1936. In Warner's first full decade in Burbank, the studio added nine new soundstages, the Mill building, and the backlot area that became Midwest Street. The studio literally raised the roof on Stage 7 (now Stage 16) and survived a devastating fire in 1934. Within the gates, WB produced an array of gangster films, Busby Berkeley musicals, Errol Flynn swashbucklers, lavish period biographies, and gritty social commentaries. The studio even won its first best picture Oscar in 1937 for *The Life of Emile Zola*. All these accomplishments took place during the Great Depression, when the studio saw losses of as much as $14 million in 1932. By the time of this 1936 photograph, much of the studio's current form was in place. Notice the reconfigured layout of Olive Avenue in the foreground (in contrast to the top image on page 35), which was curved to connect to Barham Boulevard during the decade.

CONFESSIONS OF A NAZI SPY, 1939. After a Warner office in Germany was smashed and an employee killed by a group of anti-Semites, the brothers produced a propaganda spy-thriller called *Confessions of a Nazi Spy* (1939) to alert the country to the threat of Hitler and the Nazis. For this controversial decision, the Warners received several death threats and were warned that during the film's premiere, the theater would be bombed. The isolationist U.S. Congress called Harry Warner to Washington to address charges of "creating hysteria." All was forgotten a few months later when Hitler invaded Poland. The studio re-released the film the following year with footage added of the Nazi blitzkrieg across Europe. As with sound, the Warners had once again foreseen the next big thing coming before the rest of Hollywood. But this time, they would gladly have been proven wrong.

Three

HERE'S LOOKING AT YOU

WB IN THE 1940S

WARNER FIRST NATIONAL STUDIOS, 1940. During World War II, the studio put its motto, "combining good citizenship with good picture making," into full effect in war films and bond drives, and nearly a quarter of the studio's personnel served in uniform. The labor struggles of the postwar era would teach the moguls that their word on the lot would no longer go unchallenged.

Warner First National Studios (Facing East), 1940. This photograph from 1940 was taken over an area situated between Warner Bros. and Universal Studios. Olive Avenue heads northwest to the area of "beautiful downtown Burbank" in the distance. Notice the partially paved Los Angeles River in the top right corner. Today the entire river is a concrete-lined channel. Before its completion, the bridge in the bottom right corner washed out many times.

Barham Boulevard, 1947. This photograph shows the outside of the property department and today's Stage 1 along Barham Boulevard (now Olive Avenue). The Barham Gate (now Gate 2) is located in the space between the two buildings. The "First National Pictures" portion of the name would last for another decade before being retired.

ARSENIC AND OLD LACE, 1941.
Cary Grant and Priscilla Lane
walk the lot during the filming
of *Arsenic and Old Lace* (1944).
Warner Bros. made the film while
the play was still on Broadway and
could only release it after its stage
run ended in 1944. Grant, who
reportedly hated his performance
in the film, presumably did not
feel it was worth the wait.

AFFECTIONATELY YOURS, 1941. The
stars of *Affectionately Yours* (1941)
stroll past studio soundstages. They
are, from left to right, Dennis
Morgan, Merle Oberon, Rita
Hayworth, and Ralph Bellamy. A
comedy released six months prior
to Pearl Harbor, *Affectionately
Yours* was among the lighter fare
that was squeezed off the screen
for a time by war pictures.

CASABLANCA, 1942. Great conflicts can sometime bring out the best in people, and in movies. *Casablanca* (1942) is perhaps the greatest film that Warner Bros. (or the rest of Hollywood, for that matter) ever produced. According to conventional Hollywood wisdom, this film should never have been a success. It was based on a play that was never produced, the writers were literally finishing scenes hours before filming, and at the time, Humphrey Bogart had never played a leading romantic role. But somehow it all came together and is now considered by many to be Hollywood's supreme achievement. This *Casablanca* bazaar scene was filmed on the Warner Bros. backlot, possibly on today's French Street, the same area used for the Paris café exteriors.

PASSAGE TO MARSEILLE, 1944. The cast of *Passage to Marseille* (1944) marches between a row of Warner Bros. soundstages. "From left to right are Peter Lorre, Humphrey Bogart, Charles La Torre, George Tobias, Sydney Greenstreet, Philip Dorn, Helmut Dantine, Claude Rains, and Victor Francen. Warner Bros. hoped lightning would strike twice with this film by reuniting much of the principal cast of *Casablanca* (1942), along with *Casablanca* director Michael Curtiz and producer Hal B. Wallis. During the latter years of the war, the studio continued to churn out patriotic war pictures, often at a loss, even though audiences had started to cool on them. The studio also contributed to the war effort in other ways. It sold nearly $20 million worth of bonds during the conflict, and studio personnel donated 5,200 pints of plasma. Over 760 studio employees, including Jack Warner and his son, Jack Warner Jr., served in uniform.

WARNER BROS. STARS, 1949. In this photograph, from left to right, are Barbara Stanwyck, Errol Flynn, Alexis Smith, and director Michael Curtiz. Stanwyck had one of the greatest careers in Hollywood history and was famous for her graciousness with everyone on the set, whether a costar or a crewmember. Alexis Smith was a Canadian-born actress who was often paired with Errol Flynn in WB films in the 1940s.

GARY COOPER AND EDWARD G. ROBINSON, 1949. Gary Cooper (left) made two films on the lot during 1949, *The Fountainhead* and *Task Force*. He was no stranger to Warner Bros., having won his first best actor Oscar in 1941 at the studio in the title role of *Sergeant York*. Robinson had been appearing in WB films for 15 years at the time of this photograph.

GARY COOPER, 1949. Gary Cooper drives down First Street during the making of *The Fountainhead* (1949). The monument behind his car, topped with a replica of the flag raising at Iwo Jima, was dedicated to WB employees who served in World War II. Among the names listed on the monument are those of Jack Warner and Ronald Reagan. The monument still stands near the Steven J. Ross Plaza.

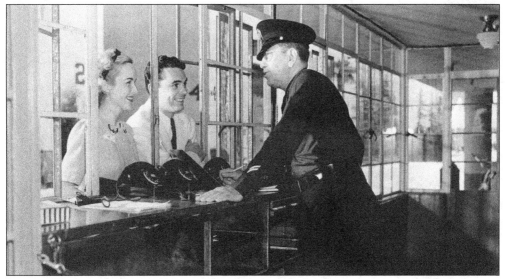

BARHAM GATE, 1946. A studio security guard chats with two visitors at the Barham Gate in the mid-1940s. Today there is still a guard gate at this site (now known as Gate 2), and the main security department for the studio is located adjacent to it. The buildings behind the visitors' heads are Stage 22 (now Stage 1) and Stage 4 (now Stage 5).

BOGIE ON A BIKE, 1940s. Even the stars need a pass! In the photograph at left from 1948, Humphrey Bogart (center) and Gig Young (right) look none too pleased to have to show their credentials to a guard to enter the studio. (One would think that Bogart's face would be familiar by 1948!) Young got his stage name from a character he played in the 1942 WB film *The Gay Sisters*. In 1955, he became the host for *Warner Bros. Presents*, the studio's first television program. In the photograph below, Bogart enters his dressing room, while his bicycle stays parked outside.

McCrea and Massey, 1940s.
Joel McCrea (above) enjoyed a
Hollywood career that lasted from
the silent era to the 1970s. By 1949,
he was starring almost exclusively
in Westerns and is seen here on a
break during the filming of *South of
St. Louis* that year. McCrea made
millions from his motion picture
career and by investing in real estate.
He married only once and lived
with his wife for 57 years until his
death in 1990. Canadian Raymond
Massey (right) descended from the
Massey side of the famous Massey-
Ferguson farm equipment empire. In
this photograph, he rides something
much smaller than a tractor to the
commissary while on a break from
filming *Desperate Journey* (1942).

WARNER BROS. COMMISSARY, EARLY 1940S. While it is not common to see world-famous performers on an average lunch hour, it happens with regularity at a movie studio lunchroom. The Warner Bros. commissary, seen above in 1940, has been located at the intersection of Avenue D and First Street (now Ashley Boulevard) for decades. The building actually has two dining areas, the regular dining room (seen in the photograph below from 1943) and a more exclusive restaurant next door, once known as the "Green Room."

Bette Davis, 1943. In this photograph from 1943, Bette Davis and an unidentified man dine together in the studio's exclusive Green Room. Few actors in Hollywood threw themselves more valiantly behind the war effort than Bette Davis. She once personally sold $2 million worth of war bonds in only two days, and along with some other A-list friends in 1942, she transformed an old nightclub into the "Hollywood Canteen," a club for men in uniform. She made it her mission to ensure that it was staffed nightly by Hollywood stars who would entertain the fighting men on leave. Two years later, art imitated life when she played herself in a fictionalized account of the club called *Hollywood Canteen* (1944). She was later quoted as saying that the founding of the club was one of her proudest achievements.

RONALD REAGAN AND PETER LORRE, EARLY 1940S. Long before his presidency, Ronald Reagan was a star on the Warner Bros. lot, appearing in over 40 films. His favorite WB film was *King's Row* (1942), which would probably have made Reagan a major star (and subsequently changed the course of American politics in the process) had he not enlisted in the U.S. Army two months later, halting his career's momentum. In the photograph above, he is seen chatting with Mary Maguire (center) and Beverly Roberts in 1940. In the photograph below, Peter Lorre dines with an unidentified man in the Green Room during the filming of *Passage to Marseille* in 1944.

ANN SHERIDAN AND JANE WYMAN, MID-1940s. Ann Sheridan (left) and Jane Wyman light up the room during the making of *One More Tomorrow* in 1946. Ann Sheridan was born in Texas in 1915, and after winning a part in a Hollywood film as a prize in a beauty contest, she acted in small parts under her real name (Clara), which she changed when her career blossomed. She made several films on the lot during the 1930s and 1940s opposite many of the studio's chief leading men. Wyman has the distinction of being the first person to win an Oscar for a non-speaking role in the sound era (in the role of a deaf-mute in 1948's *Johnny Belinda*). She would also have been first lady had she stayed married to Sheridan's *King's Row* costar Ronald Reagan, who was her husband from 1940 to 1948.

WB COMMISSARY, 1940s. Henry Fonda (left) enjoys a meal during the making of the film *Jezebel* (1938). Fonda had been in films for three years at this time and was two years away from playing the seminal role of Tom Joad in *The Grapes of Wrath* (1940). *Jezebel* would win Bette Davis an Oscar. When the statuette came up for auction decades later, it was purchased by director Steven Spielberg and returned to the Academy of Motion Pictures Arts and Sciences. In the photograph below, from left to right, are Raymond Massey, director Frank Capra, an unidentified man, Peter Lorre, and Cary Grant at lunch in 1944 during the making of *Arsenic and Old Lace* (1944).

JAMES CAGNEY, 1943. On screen and off, James Cagney was a tough guy. Jack Warner called him a "professional againster" for his refusal to back down from any quarrel. In this photograph from 1945, Cagney (center) speaks with actor Walter Huston (left) in the studio's Green Room, while an unidentified man looks on.

NEW YORK STREET, 1941. This is a view of the New York Street portion of the backlot, looking northwest from an area between today's French Street and Embassy Court. Notice the Greek Revivalist structure at the far left, which once served as an exterior for the television series *Night Court* (1984–1992) and as the commissioner's office in the campy 1960's series *Batman* (1966–1968).

BROWNSTONE STREET, 1940. Brownstone Street was first constructed in 1929 and is the oldest backlot area on the studio grounds. These block-long facades along what was once First Street (now Ashley Boulevard) have been employed any time a production called for a Manhattan street scene. The above view looks west towards what is today the Steven J. Ross Plaza, a pedestrian walkway no longer open to cars. The recessed area just past the car on the left is the location of the studio's commissary. The photograph below shows the same street looking in the opposite direction along Ashley Boulevard. The facades on the left side of the street have since been replaced by the 516-seat Steven J. Ross Theater.

SAM WANAMAKER, 1947. One of the darkest periods in Hollywood's history was the "Red Scare" of the McCarthy era. Warner Bros. actor Sam Wanamaker, seen here on the lot in 1947 during the filming of *My Girl Tisa* (1948), was in England when he learned he had been blacklisted for harboring communist sympathies. He chose to continue his career in England and was instrumental in the rebuilding of Shakespeare's Globe Theatre in London. (He is coauthor Marc Wanamaker's uncle.)

BROWNSTONE STREET, 1940. This photograph of Brownstone Street was taken a few months after and from a slightly different angle than the picture at the bottom of page 100. Note the newly constructed Stage 21 in the distance. It was built for the Errol Flynn film *The Sea Hawk* (1940). Stage 21 was designed for water filming and was built over a specially constructed pool. Ships of up to 165 feet could float on this artificial "sea."

THE SEA HAWK, 1940. Errol Flynn's seaborne swashbuckler *The Sea Hawk (1940)* could not be made until a large enough stage could be found. Since one did not exist anywhere, Warner Bros. decided to build it themselves. It was here, near today's Hennesy Street, that this huge edifice was born. At the time, it was the largest and most modern stage in the world. In the photograph at left, work crews put finishing touches on the structure. Notice the railroad car peeking from between the buildings. At one time, WB had a full-sized railroad depot here with several railcars and locomotives. The image below from the film shows one of the ships, which took 375 men 11 weeks to construct. It was built on hydraulic jacks to enable it to roll with the waves, which were created by a special machine.

THE SEA HAWK STAGE, 1940s. The photograph at right shows the construction of one of the two ships that were created for *The Sea Hawk* (1940). Workers built a 135-foot, full-scale man-of-war and a 165-foot Spanish galleass for the film, which were placed side-by-side with a 12-foot pit in between. The pit allowed stuntmen to jump into the water from off the ship, as the rest of the "sea" was only 4 feet deep. Stage 21 could also be used for dry productions, as seen below in the musical *This Is the Army* (1943). Notice the *Sea Hawk* ship in dry dock on the side of the building. Stage 21 would remain in use until being leveled after the May 1952 studio fire.

WB BACKLOT, 1940s. The backlots once stretched from Warner Boulevard on the north side of the lot all the way to the banks of the Los Angeles River to the south. Much was destroyed in the fires of 1952, while other parts were razed for new buildings. Today's backlot, which consists of Midwest Street, French Street, Brownstone Street, Embassy Court, New York Street, Hennesy Street, Downtown Plaza, and the jungle set, are just a fraction of the lots that once existed. The exterior set in the photograph above was known as English Street, and the set in the picture below was called New England Street. The proximity of the mountains indicates that this set was located near the site of the present-day Bridge Building.

WB BACKLOT, 1941. Pictured here are views of two more of the studio's backlots from 1941. The photograph above looks down a street on one of the lot's Western sets. Below is a view of a set that was frequently used for shooting scenes of small towns. At one time, Warner Bros. had three Western sets on the backlot: Western Street, Mexican Street, and Laramie Street. The last of these, Laramie Street, was replaced in 2004 with a row of New England–style homes that are used as production offices and exteriors. The new area is now known as "Warner Village."

PROPERTY DEPARTMENT, 1940. The property department contains the thousands of chairs, desks, tables, lamps, appliances, beds, paintings, carpets, draperies, and fixtures (including lots of kitchen sinks) used to dress sets to any time, place, or style. These two photographs from the 1940s show some of the items found among the 247,000 square feet of storage space that the department utilizes. Fauxmasterpieces grace the walls in the photograph above, and shelves of lamps can be seen in the view at left. The property department, drapery and upholstery department, and security office are located in Building 30, alongside Olive Avenue.

COSTUME DEPARTMENT, EARLY 1940s. Today the studio's costume department has 60,000 square feet of workspace and storage area to house 30,000 pairs of shoes, 10,000 hats, and 50,000 pieces of jewelry. Their current offices are located near the Bridge Building in the lower levels of the parking structure known as Lot A. In the 1940 photograph above, seamstresses sew gowns for an upcoming film. In the 1941 image below, a limbless army of fitting forms used for actresses Mary Brodell, Pat Lane, Joan Leslie, Ann Sheridan, Mary Astor, Bette Davis, Merle Oberon, May Robson, and Olivia de Havilland is on display.

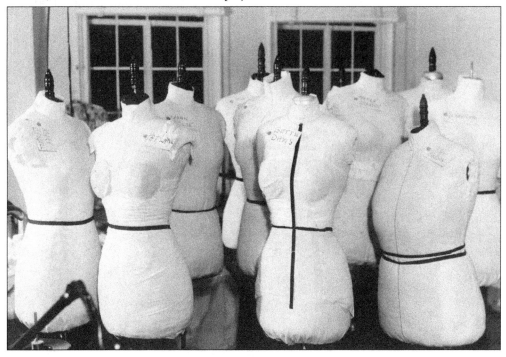

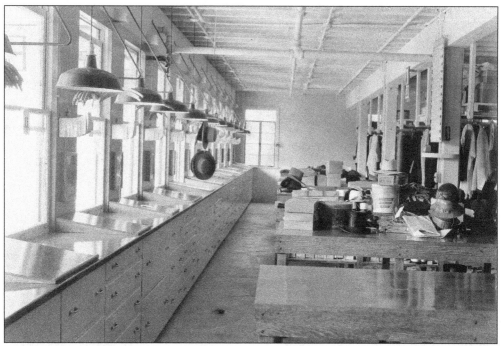

COSTUME DEPARTMENT, 1941 AND 1937. The photograph above is a look at the inside of the costume department from 1941. This area was where male extras were assigned their costumes. The photograph below from four years earlier shows the outside view of this area. Note the signage listing the productions associated with each window. On this day, windows 11–14 were being used for the William Keighley company production of *The Prince and the Pauper* (1937). At the time of these shots, the costume/wardrobe department was located near the main administration building. Today it can be found in the lower levels of parking structure Lot A.

THE SEA HAWK, EARLY 1940S. Above is a castle set used for Errol Flynn's *The Sea Hawk* (1940) on the studio's Dijon Street backlot. The photograph below shows the construction of another lot during this same time frame. Dijon Street was once located near today's Hennesy Street (also known as Tenement Street), named in honor of art director Dale Hennesy. Some of the backlot areas of the past were known as Madison Avenue, Doonevan Flats, and Bonnyfeather, Wimpole, English, New Orleans, German, and Canadian Streets. There was once even a train depot on the lot with working locomotives.

VIRGINIA CITY, 1940. *Virginia City* (1940), starring Errol Flynn, Randolph Scott, and Miriam Hopkins, was released in the period between Flynn's *Dodge City* (1939) and *The Sea Hawk* (1940). It used outdoor settings in Arizona and Vasquez Rocks, 30 miles to the north of Burbank. For this scene, set designers were able to suspend disbelief to re-create the Old West inside a soundstage, complete with horses, wagons, and a Southwestern landscape. Humphrey Bogart, who played a mustachioed gunman in this film, was on the cusp of superstardom. He would have his breakthrough role the following year in *High Sierra* (1941), which he would quickly follow up with the timeless classics *The Maltese Falcon* (1941) and *Casablanca* (1942). Incidentally, the 43-pound original Maltese falcon statuette can still be seen at the studio alongside the tiny 58-key piano from *Casablanca* in the Warner Bros. Museum.

BUILDING 5, 1940. The photograph above shows a view of First Street looking to the east. The two-story building on the right was known as Building 5 at the time (now Building 8) and was constructed at the north end of what is today Stages 4 and 11. This building served as dressing rooms in 1940, and today is primarily used for production offices. In the distance is the huge Stage 21, which no longer exists. The photograph below shows the same building from further to the right, with the bungalow area visible across the street. Building 5 appeared in *A Star is Born* in 1954.

HOLLYWOOD BLACK FRIDAY, 1945. Labor disputes were rare during the war years, as they were looked on as unpatriotic. After the conflict, Warner Bros. made record profits but was not in any hurry to share them with workers returning from overseas. The workers responded by walking off the job. The dispute turned violent on Friday, October 5, 1945, when striking workers rioted outside the Barham Gate (now Gate 2). In the ensuing melee, fights broke out between strikers and strikebreakers, cars were overturned, and police used tear gas and fire hoses to disperse the rioters. The day was called "Hollywood Black Friday" and led to the passage of the Taft-Hartley Act, which limited the power of unions to strike. It was later written about in Thomas Pynchon's novel *Vineland*.

A STOLEN LIFE, 1945. Bette Davis and Glenn Ford starred in the drama A *Stolen Life* (1946). In the film, Davis plays twin sisters in love with the same sailor. Some of the film's scenes were shot here on the studio's fishing town backlot, which was lost to the 1952 studio fires.

STAGE 5, 1945. This photograph shows the northeast corner of Stage 5 (now Stage 15) at the intersection of Third Street and Avenue D. The art department was built on the north side of the building to take advantage of the sunlight. Notice the flared supports on the outside of Stage 7 (now Stage 16). These were required for additional support after the roof was raised 35 feet in the mid-1930s.

WARNER BROS. STUDIOS (AERIAL PHOTOGRAPH), 1949. This photograph of the studio facing southeast was taken in April 1949. The changes on the lot in the tumultuous 1940s included the construction of three new soundstages, a huge water stage, and a residential district just south of Midwest Street for the film *King's Row* (1942). The mountainous, wooded area on the right side of the photograph is Griffith Park, a rugged 4,200-acre recreational area within the city of Los Angeles. Its proximity to the motion picture industry has made Griffith Park a frequent filming location. The partially cleared area at the top of the photograph was the original site of Universal Studios before it moved a few miles west to its current home. It later became Lasky Ranch and was used for the climactic battle scenes in D. W. Griffith's *The Birth of a Nation* (1915). Today it is the site of Forest Lawn Hollywood Hills Cemetery, where in 2009 a private funeral for Michael Jackson was held before his body was transported to Los Angeles's Staples Center for the public memorial.

Four

FROM THE THEATER TO THE LIVING ROOM

THE 1950s

DORIS DAY, 1951. Doris Day, born Doris Mary Ann von Kappelhoff, made her Warner Bros. debut in *Romance on the High Seas* (1948). She continued starring on the lot during the early 1950s in light musicals and the occasional dramatic role. In this shot from 1951, Day makes her way onto a stage during the making of *Lullaby of Broadway* (1951).

WARNER BROS. STUDIOS, 1951. By the dawn of the 1950s, studio moguls feared that television, the newest kid on the media block, would keep people in their living rooms instead of in the theaters. In 1946, there were only 6,000 television sets in America, but a mere four years later, that number had increased one-thousandfold to six million. Conversely, the big producers sold 100 million movie tickets in 1946 but saw that number cut in half over the next decade. Studio heads hoped they could re-excite moviegoers with grittier films following a Supreme Court decision in 1952 that relaxed the restrictive Motion Picture Production Codes. Most studios, including Warner Bros., resorted to technical gimmickry to lure audiences back, but ticket sales continued to drop. By 1955, the Warners cautiously dipped their toes into the television waters with a program called *Warner Bros. Presents*. Soon, they would produce several Westerns for television, and the medium they had feared so greatly would become a vital part of the studio's strength.

WESTERN TOWN, MID-1950S. In 1955, Warner Bros. Television was born with the debut of *Warner Bros. Presents*. Several Western adventures followed, including *Cheyenne, Maverick, Colt 45*, and *Bronco*. The studio still made Westerns for the big screen as well. The photograph above shows a scene being rehearsed in *The Left Handed Gun* (1958), starring Paul Newman (center). In the film, Newman plays Billy the Kid, who was thought to be left-handed (historians now know this was not the case). In the photograph below, the Western backlot is being prepared for a scene from *Shoot-out at Medicine Bend* (1957), which starred Randolph Scott, Angie Dickinson, and James Garner. Garner began his run on television as Bret Maverick in the popular WB Western *Maverick* (1957–1960) that same year.

SOUND ALLEY, 1955. Sound production has been a crucial ingredient of Warner Bros. filmmaking since Sam Warner started making Vitaphone shorts in the 1920s. This area of the studio by the entrance to the Warner Bros. Museum is known as "Sound Alley" because of the sound production departments that occupy these buildings. The bridge in the distance was crossed by Judy Garland in *A Star Is Born* (1954). The Eastwood Scoring Stage, large enough to accommodate a 120-piece orchestra, is at the far end of the street. This stage was refurbished and renamed in 1999 after Clint Eastwood, who is an accomplished musician. Eastwood is a longtime resident of the studio, and his *Malpaso* Productions offices occupy the northwest corner of the lot. (*Malpaso* means "bad step" in Spanish, and Eastwood was once told that starring in the "Spaghetti Westerns" that made him an international star would be a "bad step" in his career. A lover of irony, he later decided to call his new production company Malpaso Productions.)

THE GIRL HE LEFT BEHIND, 1956. Natalie Wood and Tab Hunter are pictured here in a scene from the 1956 drama *The Girl He Left Behind*. In the film, Hunter plays a young man who joins the army after flunking out of college. Much of it was shot at Fort Ord, California, but this scene was captured in front of the building that then served as the studio's wardrobe department.

DIMITRI TIOMKIN, 1953. Master of suspense Alfred Hitchcock directed several films for Warner Bros. in the early 1950s, including *Stage Fright* (1950), *Strangers on a Train* (1951), *I Confess* (1953), and *Dial M For Murder* (1954). Dimitri Tiomkin often served as the master's musical director during this period. Here he is seen conducting an orchestra for a film's soundtrack. Today this building is known as the Eastman Scoring Stage..

WATER TANK, 1950s. Since the earliest days of Hollywood, special effects artists have been fooling audiences into believing what they are seeing. In this shot, technicians employ a water tank with a huge screen behind to make audiences think they are seeing an actual ship sailing on the horizon. While it may look real on screen, notice that this "ocean" only comes up to the man's waist in the pool.

BARHAM GATE, 1959. Here is a view of the Barham Gate (now Gate 2) from the close of the 1950s, showing a bit more foliage than in previous years. During this decade, countless actors, producers, directors, technicians, extras, and regular studio workers passed through these gates daily. Notice the facing WB crest signs on the sides of the buildings that were illuminated during these years.

STUDIO FIRES, MAY AND JULY 1952. On May 16, 1952, a devastating fire swept through the backlot of the studio, turning several facades into ash. In the photograph above, workers douse sets ahead of the flames, hoping to limit the destruction. A second fire that year in early July destroyed much of what was left of the backlots. Flames leap skyward from the false fronts in the photograph below during the second fire. Actors Burt Lancaster, Gordon MacRae, Ray Bolger, and Steve Cochran were pressed into service to help firefighters battle the blaze. Between the two fires that year, 8 acres' worth of exterior sets were ruined.

STUDIO FIRE (AERIAL PHOTOGRAPH), JULY 1952. The backlots once covered almost half of the real estate on the Burbank lot. After the fires of 1952, some of the destroyed areas were rebuilt, while much of the rest was lost to the construction of new buildings. The studio had suffered a similar fire in 1934 that destroyed dozens of irreplaceable films. Studio backlots, which are frequently the sites of heavy electrical equipment and pyrotechnics, are especially vulnerable to blazes. Warner Bros. and its neighbor, Universal Studios, have the added risk of wildfires to worry about, as they both border untamed mountainous areas. These two aerial photographs were taken on July 9, 1952.

ADMINISTRATION BUILDING, 1959. By 1959, WB had been headquartered on the Burbank lot for over 30 years. During these three decades at 4000 Warner Boulevard, the studio made hundreds of musicals, gangster films, thrillers, comedies, Westerns, romances, war films, biographies, swashbucklers, and cartoons for the silver screen. In the coming decades, WB would be doing it all again and again, only now for both the silver screen *and* for television.

THE SILVER CHALICE, 1954. *The Silver Chalice* (1954), a film set in the Roman Empire, gave Paul Newman his first Hollywood role. In this photograph from 1954, Newman and the film's star, Virginia Mayo, are having test shots taken. Newman was not proud of the film (and had apparently never seen an Ed Wood movie), calling *The Silver Chalice* "the worst motion picture produced during the 1950s."

JAMES DEAN, 1955. James Dean was driving his new race car to a track in Salinas, California, on September 30, 1955, when he was killed in a near head-on automobile accident in Cholame, California. It is hard to now imagine, but at the time of his crash he was still relatively unknown to the general public, having starred on screen in only one motion picture, *East of Eden* (1955). (*Rebel Without a Cause* (1955) and *Giant* (1956) would not be released until after his death.) Ernie Tripke and Ron Nelson, the young California Highway Patrol officers who investigated his accident had never heard of him at the time. Only in death would Dean reach superstar status. He is seen here in front of the lot's Midwest Street church during the filming of *East of Eden*, for which he was nominated for a best actor Academy Award.

GIANT LUNCHEON, 1955. Director George Stevens hosted a luncheon at the commissary during the filming of *Giant* (1956). In attendance that day, from left to right, were Henry Ginsberg, George Stevens, Elizabeth Taylor, Rock Hudson, Mort Blumenstock, and composer Dimitri Tiomkin. The film was nominated for 10 Academy Awards, including best actor nominations for both Rock Hudson and James Dean (posthumously), and won a best director statuette for George Stevens. This was the third and final film of James Dean's short career. A few days after the film's completion, he celebrated the signing of a 10-picture contract by purchasing a Porsche Spyder, which he intended to race in Salinas the following day. The intersection of California Highways 41 and 46, where Dean died in a car crash, is now known as the "James Dean Memorial Junction." Nick Adams, who could mimic his friend Dean's voice, was called on to overdub some lines for the film after the crash. *Giant* was Warner's highest grossing film of the 1950s.

STAGE 4, 1956. Stage 4 (now Stage 5), near today's Gate 2 entrance, housed its share of landmark films during the 1950s, including *Dial M for Murder* (1954), *Giant* (1956), *The Spirit of St. Louis* (1957), and *Rio Bravo* (1959). The building to the left contains today's Stages 1, 2, and 3.

THE BRAMBLE BUSH, 1959. The studio's New York Street has been re-dressed to look like a building in Cape Cod, Massachusetts, for the filming of *The Bramble Bush* (1960), starring Richard Burton and Barbara Rush. New York Street was originally built in 1930, but a studio fire destroyed most of it in 1952. Today's New York Street was re-created atop the ashes of the original structures.

WARNER BROS. WATER TOWER, 1959. During the 1950s, Warner Bros. Studios churned out several classic films, including *A Streetcar Named Desire* (1951), *A Star Is Born* (1954), *East of Eden* (1955), and *The Searchers* (1956). It also saw the emergence of Doris Day and the return of the Warner Bros. musical to the screen. During this decade, the Burbank lot suffered two major fires, which reconfigured its backlot, and a new jungle set was added, complete with a 250,000-gallon lagoon. Change was everywhere on the lot during these years, especially at the top. The close of the decade brought the studio something unique in its history—a single Warner brother at the helm. The company name would remain in the plural, but never again would more than one Warner brother steer the media empire that sprouted up from a single nickelodeon in Pennsylvania.

Visit us at
arcadiapublishing.com

CPSIA information can be obtained
at www.ICGtesting.com
Printed in the USA
BVHW011140030320
573945BV00007B/19